"Angela's book is just as important for personal brands as it is for companies. In this new world of work, everybody must understand how to facilitate and grow a community, in order to achieve maximum success. Angela's book will guide you to community manager nirvana!"
Dan Schawbel, Author or 'Me 2.0: Build a Powerful Brand to Achieve Career Success'

"Journalism is no longer a lecture; it's a conversation with and among the audience. Engaging readers and viewers to engage in online communities is an essential part of creating modern media. Angela's book provides useful, actionable information about how to build and nurture online communities based on real-world experience."
Mark Potts, Co-Founder of
http://www.WashingtonPost.com,
http://www.Backfence.com **and author of**
'http://www.RecoveringJournalist.com.'

"Angela begins the pioneering task of setting the rules for online communities in this must-read book. Her sass, wit and sheer knowledge of this unknown frontier are great guides for anyone wanting to enter the online community space."
Maren Hogan, Principal, Red Branch Media

"Being able to attract and manage over 11,000 members proves that you're an expert when it comes to community engagement. In this book, Angela Connor not only shares her own experience, but includes the opinions and ideas of other community practitioners. The result is a book that should be considered required reading for anyone involved or interested in the art of community building."
Martin Reed, Community Developer/Manager,
http://www.CommunitySpark.com

"A very conversational, wonderfully written, action-oriented, read with excellent examples."
Janet Clarey, Analyst & Sr. Researcher, Brandon-Hall Research

D1316564

"In an era of rapid-fire change, Angela understands that Community is a slow-burn enterprise. She has created a personable, practical primer for those individuals and companies interested in enabling connectivity and exchange."
Venessa Paech, Community Manager,
http://www.LonelyPlanet.com

"Angela tells you the score on running an online community with verve and humor. She knows what she's talking about, and if you run an online community or want to, you should listen."
Lisa Williams, Founder and CEO,
http://www.Placeblogger.com

"In 2009, savvy public relations and marketing professionals are honing in on the importance of connecting with targeted, niche online communities. Angela pulls on expert insight from thought leaders across the social Web to provide an easy-to-digest slate of guidelines to remind us all of what it takes to connect effectively with target audiences. A crucial read for any social media newbie looking to learn the online community rules of the road."
Scott Meis, Sr. Project & Social Media Director, Carolyn Grisko & Associates Inc.

"Angela lays out some great points on community engagement with real life examples that give readers the how-to when implementing these strategies within their own business. Not to mention, it's all written in a simple to read manner."
Sonny Gill, Social Media Strategist,
http://www.SonnyGill.com

"The new journalism is becoming less a "telling" of stories and more a conversation with our communities. But how do we move from the old model of circulating the news to the new model of managing these social groups? Angela Connor provides a clear and concise map to follow, whether working from the corner office or a corner of a coffee shop."
Ron Sylvester, Interactive News Reporter,
http://www.TheWichitaEagle/Kansas.com

18 Rules of Community Engagement

A Guide for Building Relationships and
Connecting With Customers Online

Angela Connor

20660 Stevens Creek Blvd., Suite 210
Cupertino, CA 95014

First Printing: May 2009
Paperback ISBN: 978-1-60005-142-5 (1-60005-142-1)
Place of Publication: Silicon Valley, California USA
Paperback Library of Congress Number: 2009923129

eBook ISBN: 978-1-60005-143-2 (1-60005-143-X)

Trademarks

Warning and Disclaimer

A Message from Happy About®

Thank you for your purchase of this Happy About book. It is available online at http://www.happyabout.info/community-engagement.php or at other online and physical bookstores.

- Please contact us for quantity discounts at sales@happyabout.info
- If you want to be informed by e-mail of upcoming Happy About® books, please e-mail bookupdate@happyabout.info

Happy About is interested in you if you are an author who would like to submit a non-fiction book proposal or a corporation that would like to have a book written for you. Please contact us by e-mail editorial@happyabout.info or phone (1-408-257-3000).

Other Happy About books available include:

- I'm on LinkedIn—Now What (second edition)???
  http://www.happyabout.info/linkedinhelp.php
- 42 Rules™ for 24-Hour Success on LinkedIn
  http://www.happyabout.info/42rules/24hr-success-linkedin.php
- I'm on Facebook—Now What???
  http://www.happyabout.info/facebook.php
- Twitter Means Business
  http://www.happyabout.info/twitter/tweet2success.php
- 42 Rules™ of Social Media for Business
  http://www.happyabout.info/42rules/social-media-business.php
- Happy About an Extra Hour Every Day
  http://www.happyabout.info/an-extra-hour.php
- The Successful Introvert
  http://www.happyabout.info/thesuccessfulintrovert.php
- Communicating the American Way
  http://www.happyabout.info/communicating-american-way.php
- The Emergence of The Relationship Economy
  http://www.happyabout.info/RelationshipEconomy.php
- 42 Rules™ for Driving Success With Books
  http://www.happyabout.info/42rules/books-drive-success.php
- Rule #1: Stop Talking! A Guide to Listening
  http://www.happyabout.info/listenerspress/stoptalking.php
- 42 Rules™ for Successful Collaboration
  http://www.happyabout.info/42rules/successful-collaboration.php

Dedication

To my darling daughters, Kalyse and Kaiya, for their patience, and my husband, Derrick, for his love and support.

Acknowledgments

First and foremost, I have to thank my Twitter pal and fellow social media enthusiast Bryan Person for connecting me with Happy About founder Mitchell Levy, who made a commitment to this project just a few short days after our initial phone conversation. Mitchell understood and believed in my vision. I am grateful for everyone who took an early sneak peek of the book and stood behind it with their endorsements. You gave me the confidence I needed to keep pushing. To my partner-in-crime and developer extraordinaire, Rusty Kroboth, for giving me all the tools I need whenever I need them and to the GOLO community, for giving me enough great days to make it through the tough ones. A special thanks to photographer Pamela Mullins for my photo on the back cover, and to those who consistently read my blog, Online Community Strategist, and share their expertise in the social media space. And many thanks to the visionaries at WRAL.com who let me carve my own niche.

Acknowledgments

Contents

Foreword by Peter Shankman

We live in a world populated by more requests for attention than at any other time in history. From the time our alarm wakes us up in the morning, until the time we go to sleep, we're assaulted by a non-stop barrage of requests for attention. Twitter this -- Facebook that -- Advertise the other thing -- Join this community, be a part of this group -- it's beyond overwhelming -- it borders on obscene. And if you're tasked with managing it all for your company...best of luck! There used to be a commercial that talked about "herding cats." That's very 2002 -- this is all about herding millions of people -- every day, and forever. How do you get them there? How do you make them "come to your side," as it were?

Back in the mid-90's, I worked for America Online -- one thing they were hardcore about was never forgetting that it's the user who drives the experience. The company makes the product, but the user has to like it -- without them, making the product is futile. That's about a million times truer now then it was back then. Now, the question is choice. If you can't find a way to engage your user, give your customer a feeling of personalization each time they shop with you, and most importantly, make them feel like they need, not just want, to come back, then you'll lose. Remember, Amazon started out as a bookstore. Are you listening? Are you building communities and truly engaging all your interactions? If not, put this book down, and go get a Latte and the latest edition of People. But if you know you need to do it, this book might just be able to show you how.

Peter Shankman
Founder, Help A Reporter Out - HARO
http://helpareporter.com

Section 1: The Basics

Once upon a time, it was all about audience. If you had a product or service to market or a message to send, you needed an audience to receive it. Whoever broadcast their message to the largest audience was the clear winner and it was very much a one-way street. But in the age of instant communications, where we are constantly connected and faced with infinite choices, consumers want and expect more.

We are now living in the conversation age, where one-way communication is no longer acceptable or desired. People want to engage and discuss, react and interact.

It is no longer effective to have an online presence without interaction.

Communities make that possible.

The fundamental difference between an audience and a community is the former can be bought, while the latter must be built.

In the first section of the book I discuss the basics of community and share insight from industry thought leaders on the characteristics and dynamics of communities, including a specific principle that characterizes those dynamics.

If you think it's insignificant, believe me, it isn't. The basics are crucial. So read through the first section to get your arms around the concept of community building and then we'll hit the rules.

1 If you build it will they come?

The answer, simply, is NO! Many organizations and businesses mistakenly believe that if they provide the tools for community engagement and interaction, a community will form on its own and ultimately engage and interact. Nothing could be further from the truth. Creating an online community or social network with user profiles, blogs, forums, chat rooms, image galleries, and other bells and whistles will not make it a destination for compelling conversation or encourage users to create content. Allowing comments on blogs and news stories won't make people post them, nor will opening a chat room attract large groups of people who will enter and start chatting. Along those same lines, creating a forum won't make interesting topics suddenly appear. Providing the tools is only the first step toward building and growing communities and it isn't the most important one. While providing the tools does indicate a desire to bring people together, it does nothing to actually make it happen.

It takes a different kind of investment to grow community, and a major portion of that investment is TIME. The other part is engagement. If you don't have the time or patience to engage and do so genuinely, or if you're unwilling to pay someone who can do it on your behalf, you cannot realistically expect to grow a community around any topic, or succeed in an existing one. What you will do is waste a lot of time and set yourself or your organization up to fail. My advice to you would be: don't even bother.

Why communities fail

A study of more than 100 businesses with online communities found that 35% had less than 100 members and less than 25% had more than 1000 members. This was published in the Business and Technology section of the *Wall Street Journal's* Web site on July 16, 2008.[1] The headline was: "Why Most Online Communities Fail." According to the article, Ed Moran, the Deloitte consultant who conducted the study, indicated that most of the sites failed to attract visitors because businesses focused on the value the community could bring rather than investing in the actual community.

That was a big mistake, and most of the people who took the time to leave comments with that story agreed. A blog poster by the name of Mitch Bishop wrote: "The success of online communities is directly related to the passion of the participants, not the money invested by the under-writer." Susan Salgy of WebWise Solutions, a company that creates corporate Web sites and Web communities wrote: "We have seen this time and time again—companies want the benefits of a community without ponying up the content and attention that will deliver the core value to community members." She went on to mention that her organizations' best clients understand the scope of the commitment, and provide the necessary long-term nurturing that will make it a success.

The key phrase in that statement is long term. Success will not happen overnight, and anything short of a long-term commitment will produce mediocre results. Communities fail when no one is tasked with providing that long-term nurturing. Communities fail when they are neglected and taken for granted and when the assumption is made that it will always exist or that if you build it they will come. Communities fail when the endurance needed for success is underestimated or misunderstood.

The recommendation made by Ed Moran, the consultant with Deloitte, was dead on: "Put someone who has experience running an online

---

1. http://www.tinyurl.com/5jahuy (http://blogs.wsj.com/biztech/2008/07/16/why-most-online-communities-fail/)

community in charge of the project." I'm convinced that this is the best solution. In fact, it's the only solution. Enter, the Community Manager.

What is a community manager?

What exactly is a community manager? And what does this person actually do? Well, it depends heavily on the goals of the individual, group, or organization behind the community. The goals of a company looking to grow brand recognition, connect with customers, and grow its customer base will differ slightly from an organization or individual interested in bringing together cancer survivors or music enthusiasts. A blogger working to build a video-gaming community will have a different set of goals and perhaps a different approach than a retail store like Pottery Barn, a cable giant like Comcast, or a nonprofit organization like the American Red Cross.

These differences make the role of a community manager very unique and underscore the importance of having clear goals and knowing what constitutes success. With GOLO, WRAL.com's online community, I strive to attract new members who live in or have strong ties to the Raleigh/Durham area (North Carolina). I want them to feel that GOLO is the best local community on the Web, where they can make friends, learn from others, and voice their opinions about the things that matter most, the great majority of them being issues that are geographically relevant. The original job description for which I applied stated the following: "Energetic, community minded person needed to oversee all aspects of content creation and editing for new community based internet product. The ME will provide vision and long range planning/direction for all content areas while managing balance between staff, freelance and community generated content related to the Raleigh/Durham area."

The job also involves cultivating relationships, and with this tagline: "Go Local. Talk Local. Share Local" it's easy for me to stay in line with the day-to-day mission and long-term strategies. That tagline guides almost everything I do within the community. Without a clear-cut mission, you will find it difficult to reach your goals. General goals such as "Reach out to the community and communicate" will only get you so far. What are you reaching out to the community for? What are you communicating about? Those are the questions that have to be answered so you can gauge your success. FreshNetworks, a European firm that builds, manages, and moderates online communities for brands such as Microsoft, HSBC, and Procter & Gamble stresses the importance of the community manager and the need to focus on the skills and strategies needed to build, grow, and manage an online community.[2]

In a call for participants for the International Online Community Management Association, German blogger Sascha Carlin describes online community management as a challenging profession that involves facilitation and moderation and refers to community managers as product managers of a special kind with a potential audience of millions. The challenge, according to Carlin, is knowing how to reach these people, what services to offer to them, and how to get them involved in our companies' business goals.[3]

Community strategist Connie Bensen characterizes the position as "broad and encompassing," with this definition: "A community manager is the voice of the company externally and the voice of the customers internally. The value lies in the community manager serving as a hub and having the ability to personally connect with the customers (humanize the company), and providing feedback to many departments internally."[4] While Bensen's definition seemingly applies to enterprise only, phrases like "personally connect" and "humanize the company" are far from corporate. They bring personality into play and that resonates across the board. The rules of engagement are the same for Ford, Comcast, and JetBlue as they are for Pottery Barn, *The New York Times*, bloggers, marketers, business professionals, and entrepreneurs. They just have to be tailored to meet individual and specific goals. Some of the 18 rules laid out in this book will be more helpful than others but each rule should be practiced at some point to determine which deliver the best results.

Beyond the "role"

More important than the role, so to speak, are the attributes of the individuals filling it. The face or voice of any community should be a committed individual who will reach out to community members, encourage them,

---

2. http://www.tinyurl.com/cwlyup (http://blog.freshnetworks.com//2008/12/lets-focus-on-how-we-build-and-manage-online-communities/)
3. http://www.iocma.org/community/blog/13-iocma-call-for-participants
4. http://www.tinyurl.com/clhb5n (http://conniebensen.com/blog/2008/07/17/community-manager-job-description/)

value them, and make sure they know their presence is appreciated on a daily basis. They will troubleshoot, sympathize, empathize, and make things happen.

If there is no one actively engaging with users, and doing so with a purpose, the community will cease to exist. That said, community managers have a tall task. So what's the most important role of a community manager? I threw that very question out to my twitter network and received several interesting answers. Martin Reed, author of the blog Community Spark[5] and creator of the online community Female Forum,[6] said the role of a community manager is to facilitate, encourage, and develop relationships. Blogger and veteran copywriter Scott Hepburn said the most important role is that of host: making introductions, announcements, and fulfilling needs. Community manager Holly Seddon stressed the importance of respect. She says you must respect members but at the same time maintain an ability to keep coherent boundaries in place. Deb Ng agrees. As community manager of Blog Talk Radio[7] and cocreator of Kommein.com, a Web site focused on community building, she feels the most important job for a community manager is to keep the lines of communication open and foster relationships.

You can't force community

In a post on the blog Branding David, author David Peralty mentions very matter-of-factly four words that anyone who has attempted to bring people together online and form communities is quite familiar with and know to be gospel: *You can't force community.*[8] You can build it, foster it, cultivate it, and shape it. You can nurture it, believe in it, and support the members who make it what it is on a daily basis. But you can't force it. Keep in mind that shared interests is what brings people in a community together, and online communities can only thrive if people visit regularly and spend a good amount of time when they do visit. And given the fact that no one willingly wastes this precious commodity, it should be a major priority to create experiences that are worthy of their time and make them want to

---

5. http://www.communityspark.com
6. http://www.femaleforum.com
7. http://www.blogtalkradio.com/
8. http://www.brandingdavid.com/business/you-cant-force-community/

return and give even more of it. In a nine-page document called the Online Community Manifesto, author Richard Millington writes about some of the things we need to know about communities: "We need to know what motivates people. We need to know the difference between intrinsic and extrinsic motivation. We need to know how to create communities founded on these motivations."[9] A good community manager will strive to learn those things about the community. Once learned, it's easy to take what you know and keep the community engaged. When you're running a voluntary ship where time is donated and can't be bought, you're left with only one option and that's to earn it. I want to help you do just that. In this book, I will share what I know and some of the things I've learned from others while managing the online community GOLO.com, from its infancy to its current status of more than 11,000 members with dozens joining every day.

---

9. http://www.tinyurl.com/c9vkop (http://richchallenge.typepad.com/files/communitybuildingmanifesto-1.pdf)

2 User participation and the 90-9-1 principle

Have you heard of the "90-9-1" principle? It essentially states that in any given community or social group 90% of the users are lurkers, 9% are contributors, and only 1% participate "very often."

There is lots of debate about this particular principle among community leaders but most agree on the concept, though they may not necessarily agree on the exact percentages. So, if only 1% of the overall community is producing most of the content, (blogs, images, comments, forums, etc.) a community manager's job is never done. This underscores the importance of having a long-term strategy and a plethora of tools in your toolkit to turn lurkers into contributors and to encourage contributors to ramp it up a bit and move into the zone of those who post "very often."

Every community manager's goal should be to shift that paradigm, not accept it as gospel and become complacent. One of the most common questions I receive via e-mail from people managing new communities is how to get people to post more content. The truth is, in a new community where there may be 100 active members, the percentages work against you as the number of users participating very often, based on the 90-9-1 principle, would be one. And that one may be you.

What community managers think of the "principle"

When I blogged about this particular topic, there was a great response from people in the trenches, and quite the debate ensued. Their comments, insight, and thoughts on the topic were much better than the original post. This comment was left by Martin Reed, author of the blog Community Spark and creator of http://www.FemaleForum.com:

> I am not a fan of the 90-9-1 'principle', as I mentioned in my last blog post. I think it is often used by those involved in communities as a way of justifying their low levels of member activity. In my opinion, if only 1% of your members are actively creating fresh content, something is wrong. It is up to the community administrator/manager to work on getting more members active, not hide behind a supposed 'principle'. Surely if 99% of people don't regularly contribute there is something inherently wrong with your community.

Patrick O'Keefe, manager of five online forums and author of the book *Managing Online Forums*,[10] calls the principle a reality, though not necessarily a reality for everyone. Here are his thoughts on how the 90-9-1 principle applies to online forums:

> 90% of total website visitors read content and rarely contribute, 9% sometimes reply to other threads and rarely start new ones and 1% regularly start their own threads. That's the principle. Breaking it into two groups, it would be more that 90% of the visitors (that is all visitors, including search engine traffic, etc.) read and don't contribute while 10% contribute at least somewhat regularly. Not that only 1% contributes regularly.

Jake McKee, a well-regarded community blogger and creator of 90-9-1 .com, a Web site billed as a single collection of description and support for the principle wrote that he is not necessarily a believer that this principle is spot on: "I created 90-9-1.com because I realized that we practitioners didn't have a single spot to point audiences, clients, and colleagues to help us explain the concept itself. I think there's also a dynamic at play here that applies to the type of community."

---

10. O'Keefe, Patrick. *Managing Online Forums* (AMACOM 2008)

Regina B's comment underscores Jake's focus on the type of community. As the person charged with managing GuildWars.com,[11] a gaming community dedicated to the award-winning Guild Wars series, she doesn't think the numbers are accurate for her community, though the underlying principle is quite sound. "In my community, I would probably further refine the 90% into sophisticated lurkers, those who visit the game's third party fan site forums, and unsophisticated lurkers, those who do not participate in forums at all and whose only interaction is purely the game."

Scott Dodds, community manager for the Lithosphere community[12] where he helps enterprises build deeper relationships with their customers, to grow brand affinity and product advocacy says these are the most useful aspects of the 90-9-1 principle:

1. If you want to increase quantity of activity in your community, it's more effective to increase the total population who visit your site than to get current members to participate more (not that you shouldn't do both, but the former will typically be more effective than the latter).

2. If you want to increase the quality of activity in your community, focus your efforts on that 1% who contribute the most.

3. If you want to find out what the total reach is of your community, be sure to count the 90% or so who are spectators as well as the 10% who are posting.

In addition to the above strategies listed, Scott maintains that what's really important is how you can use this knowledge to further the goals of your community. While there are differences between active participants, and those who mostly lurk, both groups are consuming the content within the community and witnessing the dynamic that is developing among members. You never know what will turn a lurker into a participant or make them jump in head first never to look back. So work hard to make the community the best it can be and bring your a-game on a daily basis. If you do that, the percentages won't matter much at all.

---

11. http://www.guildwars.com/
12. http://www.lithosphere.lithium.com/lithium/

User participation and the 90-9-1 principle

3 The road to engagement

The word engagement is in the title of this book because it is more important than anything if your goal is to build, grow, or successfully interact within an online community. The word "engage" is a verb and therefore requires one to act. In Webster's Dictionary it is defined as such: (1) to occupy the attention or efforts of, involve and (2) to attract and hold fast. There are many ways to attract an online audience, but keeping one requires creativity, perseverance, and a lot of work.

Think of all of the work that goes into a personal relationship and that special attention, care, and concern that is doled out in the very beginning. Now think about how quickly the relationship falls apart when it starts to wane. If you want to keep the other person engaged and vested in the relationship you have to keep up that same level of special attention, care, and concern. Communities are no different and in both cases, actions speak louder than words. We've already established that simply building it won't make them come. But if you engage them while they're there, they'll come back, and they just might bring others with them.

You've got what it takes

I recently met with two financial advisors at a local coffee shop to give them a few tips on turning their word-of-mouth marketing into social media marketing, particularly by way of online communities. I'd met one of them the previous weekend at a birthday party for a 3-year-old where he told me that he was very good at face-to-face networking and organizing events for networking purposes but was pretty much overwhelmed by all of the social media platforms and didn't really know how or where to start.

During the meeting I asked them both several questions about how people can manage their money given the state of the economy, apprehension about the recent bank bailouts, and in the midst of record foreclosures and job loss. I asked them what their clients and potential clients were most afraid of and what kinds of questions people were asking them. They shared that people are really worried about job loss and their investments and offered tons of practical advice and tips that I later characterized as extremely useful information. I told them that while it was nice that they were relaying this information in person, they were missing tremendous opportunities by not doing the same thing on the Web through online communities. I suggested that they take this expert knowledge and delve into existing local communities to establish themselves as the experts they are online, the same way they do offline and perhaps later consider building a community of their own.

This is something they had never considered and I finally convinced them that dispensing such useful information, such as how to prepare for job loss and holding on to your finances during this economic recession, would be like handing out cups of water during a drought. They left the meeting in stealth mode having been taught a few new ways to attract potential customers and broaden their reach. I will discuss the importance of providing useful information later on in the book but I am mentioning it here so that you can start to think about the information you possess that others would deem very useful and pass along to others.

Just as the community manager's role is to engage, so is the role of members and contributors. Those who do it well, reap the benefits. If you manage to establish yourself as an expert on any subject and actively contribute to an online community with useful information related to that subject, it won't go unnoticed and there will be rewards. You will build credibility in that community and for the business professional that could result in new clients, new leads, and new opportunities that would not have otherwise come to pass.

As we move into the 18 rules of community engagement, it's important to note that these methods can be used by anyone, from the community manager to the small business owner or PR professional to the newest members of the community. The bottom line is this: Give people something to talk about, engage them, and provide useful information. Those are the key aspects of growing an online community. If you want to learn how to do it, read on.

The road to engagement

Section 2: The Rules

Growing an online community for me was trial by fire and in some aspects it still is. What seems like a great idea can easily flop, and the simplest ideas can resonate with the community in ways you couldn't imagine, bringing new members in waves.

It takes an enormous amount of work. You have to have personality, tact, an amazingly thick skin and a work ethic that will not quit. You also have to genuinely like communicating with people.

There is no surefire way to make any community the ultimate destination, but I've learned that there are surefire tactics you can employ to connect with people and help them connect with others.

You have to do more than provide the tools. You must facilitate the process, engage, interact and create an environment where people feel appreciated, important and special.

I've nurtured a brand-new community from its infancy to more than 12,000 members, and I'd like to share with you how I made that happen and how you can do the same.

4 Stroke a few egos

Community managers must check their own egos at the door and realize that without the community, they have nothing. You need them more than they need you, and that will pretty much always be the case. There can be a fine line between stroking egos and coddling individual members and you will sometimes find that you have to do a little bit of both.

There's no harm in letting people know that you need them. In fact it gives them a sense of pride and ownership in the community that ultimately serves you very well. So even after you've doled out your daily compliments (which we'll discuss in Chapter 9), you still have to stroke a few egos. You have to let the major players know that you see them as major, and the prolific bloggers have to be told how well they write. When someone posts pictures of a wedding or a new baby, compliment them on their lovely family and breathtaking photography and tell them you want to see more.

If someone shares a bit of information that resonates with you, tell them as much. If their jokes leave you in stitches, encourage them to keep them coming. Conduct interviews with the pillars of your community and post them prominently for everyone to see. Reward your top posters, continuous content creators, and keepers of the community. Offer them a stake in the community with additional duties if that's something you can offer. Make them feel special and let them know you're watching. If it sounds like a huge task, that's because it is. It's an important one too and should be done on a daily basis. It isn't something you have to spend hours doing, but it's wise to carve out at least 30 minutes of your day to recognize the members who keep the community afloat, and spend a good portion of their time on your site.

In case you're wondering how you can do this effectively and efficiently, here are a few suggestions:

- **Tell them you miss them.** ("Haven't seen you in a while, I hope everything is okay. We miss your humor.")
- **Send a personal e-mail.** ("Hey, you were one of our top posters last week. Just want you to know how much I appreciate your time. Keep it up!")
- **Ask for input.** ("I'm thinking of making some changes to the chat tool, and would love your thoughts.")
- **Encourage communication with other community members.** ("Johnny23 is looking for tax advice, aren't you a big time accountant? Maybe you can help.")
- **Make a promise, and keep it.** ("If you do decide to take more pictures let me know and I'll feature them on the home page. Your photography is top-notch.")

These types of exchanges may seem insignificant but they work wonders and make people feel valued and appreciated. Whenever I receive a personal e-mail from a blogger whose blog I read and comment on frequently, it makes me feel a tad bit special and I know they want me to keep participating. I also recognize it as an effective tactic to keep me engaged because I use it myself with my own community. If you produce content on the Web and participate in any type of social medium, you must give back in some way to those who consume your content and attempt to interact with you. There is no gesture too small. It can be as simple as sending a "thank you."

Get to know the community

I've found that it's easy to stroke a few egos when you have personal knowledge about community members. As the managing editor of GOLO.com, I come in contact with a lot of members both in public and in private. Many members communicate with me openly on my profile page and others e-mail me behind the scenes. I encourage both methods and respond to all inquiries. I want the members to feel comfortable and if they prefer private communication I most certainly oblige.

I am simply amazed at how much I've learned about individual members through such methods. From reading blogs, viewing image galleries, and paying close attention to the way members interact, and by following their comments, I have somehow learned a little about a whole lot of people. I can name the professions of at least thirty members, personal hobbies of others, and even first and last names for many, including those of their children. I know that we have members with children in Iraq, members who have lost children, members who are recently divorced, recovering alcoholics, fighting eating disorders, and battling cancer.

I know that one member is a chef, another owns a landscaping business, another is obsessed with "The Rock," and another has the most disgusting feet you ever want to see. This is a new level of sharing and highly valuable information that you can use to build relationships even further. When you're properly armed with valuable information and pay attention to what's going on in the community, ego stroking comes pretty easily. I find it pretty painless, and believe it or not it can be quite rewarding.

Stroke a few egos

5 Don't be pushy

If you have a product or service that you think most people or a large percentage of the people in an online community would find interesting, I think you should delve right in and give it a shot. After all, isn't one of the goals behind marketing through online communities to essentially capitalize on the sheer numbers and niche topics? If done properly, it can be quite effective. If done poorly, as it most often is in my opinion, it can backfire in a way that can turn ugly fast. I've seen it time and time again.

A well-intentioned individual joins a community and casually starts mentioning their travel Web site or automotive services, complete with links in every post, and the promise of a discount. What often happens next is they receive a slew of comments from the natives about the community not being a place to sell their services and it's all downhill from there. If you're lucky, you'll simply be ignored. It's a delicate balance, but it can be done. It takes time and a genuine interest in reaching the community in a respectful manner and on the community's terms.

I once saw a new member upload 750 images of wristwatches. It was his entire catalog, and I wondered how he even made time to upload them all manually. Need I say what happened to him?

He was quickly branded with what I call the scarlet letter, "S." It stands for SPAM. It's a word you don't want to be associated with in an online community. I've seen people try to work their way back in with members after coming on really strong and pushing their product or service, and it isn't pretty. It's essential to have a feel for the community before jumping in head first and at least some understanding of the culture. So, before you jump right in to the next community, here are five things you shouldn't do. Remember, this isn't the "do" list, it's the "don't do" list.

- Add links to your Web site in every single blog and comment you post.
- Write blogs with titles like: "Great deals on travel" and only mention your organization. It's the quickest way to illustrate a lack of genuine interest in the community.
- Start blogging about your product or service the minute you create a profile or join a forum or group. Your lack of history will be noticed and your intentions will immediately be clear.
- Misrepresent yourself as a satisfied customer, just to convince others to get on board.
- Disrespect the culture of the community. Take time to see how things work before you jump in and shake things up.

There is one member in the community I manage who I believe has discovered how to coexist in an online community by choosing the right balance of promotion and engagement. I asked Deidre Hughey, *partner at Dancing Elephants Achievement Group of the Carolinas*,[13] to share some of her experiences in the GOLO community with me via e-mail so that I could share them at a seminar I held for the National Home Furnishings Association in High Point, North Carolina. Deidre wrote something in her first sentence that I think is very important and true: "You have to have a 'burning desire' to get your message out to the public."

I could be wrong but I attribute that sentiment to the time commitment required to gain respect in these communities. It is very important to interact within an online community and refrain from pushing your product or service every time you post. You have to understand that they don't want

---

13. http://www.deagsales.com/

you to slam your product down their throats. If you do, you will be rejected and getting back in the good graces of the community will be virtually impossible. Deidre also indicated in the e-mail that it can be tough and that she has wondered if the community is indeed her target audience. But one thing she finds valuable is the ability to fine-tune ideas by seeing if they resonate with users, stir controversy, or "fall flat."

Missed opportunities

Just as some marketers are extremely pushy, others have little to no presence in online communities outside of their own. As the editor of an online community with more than 11,000 members, I see at least five to ten really good opportunities for marketers and advertisers on a daily basis. They unfold before my very eyes and sometimes seem too good to be true. The conversations take place in blogs, on individual profiles, and even on the comments sections of stories and blogs as well as in chat rooms. I couldn't really figure out why these opportunities were being missed at first, but now I believe it has everything to do with the time it takes to join the conversations. I think it would be time well spent, particularly on a local site for a local business because some online communities are filled with thousands of local folks who could bring lots of business to those who do it the right way.

The subject of Lasik eye surgery was a hot topic on GOLO one morning after a member mentioned in a blog that he was seriously contemplating having the surgery and asked for advice. It came in droves. Two local businesses were mentioned by name and several individual doctors were recommended, again by name. There was mention of a bad experience and someone chimed in saying they too had been considering the surgery and wanted to find a doctor and clinic with a good reputation and satisfied patients. It seemed to me like the perfect opportunity to jump right in—but no one did.

Can you imagine what would have happened had anyone from those offices gotten involved in that conversation, perhaps offering a special, a consultation, expertise, or even tips on how to select a doctor? What I saw was a potential gold mine and an opportunity to recruit customers who would spread the word and bring in even more referrals. The cost: Zero. Well the real cost is time. I guess for some, that price is just too steep.

Jeremy Lindh, founder of CornHolePlayers.net[14] has always encouraged retailers to become members and actively engage in his communities. He says some take great advantage of it, and others don't want to take the time or energy to bother with it. When done correctly, he's seen members do major business through the recommendations and compliments their products receive in his forum, although a recent attempt was unsuccessful. After a member posted about a tailgating product he really enjoys, Jeremy e-mailed the company thinking they might want to capitalize on the opportunity but was completely ignored. He chalks it up to the matter of micro-marketing not being viewed as worth their time.

Bryan Person, social media evangelist at Live World, Inc,[15] a company that builds, operates, and moderates social networks and online communities offers several reasons why marketers aren't jumping into some of the conversations taking place in online communities, such as the one about Lasik eye surgery and the tailgating product.

- They (or their PR reps) aren't doing a good job of monitoring the social web, and they haven't seen this conversation.
- They are monitoring, see the mentions, and it just takes time to respond.
- They are monitoring, see the mentions, and don't know exactly how to respond.
- They are monitoring and see the mentions, but the high volume makes it impossible or unrealistic to jump into all conversations (for many brands, this is certainly the case).
- They are monitoring, following the conversations, and simply choose not to respond.

Community expert and blogger Richard Millington[16] agrees with Bryan. He says that too many companies focus on the macro sales rather than the micro sales (which is what engaging smaller, niche communities would yield.) "They don't realize that the macro attempts nearly always push people away, while the micro always draws people closer to the company."

---

14. http://www.cornholeplayers.net/
15. http://www.liveworld.com/
16. http://www.feverbee.com/

In "Strategies for Interactive Marketing"[17] Forrester research analyst Josh Bernoff wrote that social applications in particular, such as communities and social networking sites, are cost-effective and have a measurable impact on prospects' decisions in the consideration stage, which will be important to companies under recessionary pressures. He continues with this: "Interactive marketers should stop toe-dipping and invest only in programs that can deliver on measurable metrics."

When it comes to online communities, the level of investment will vary from business to business as it also varies among individuals. Online communities require a different kind of investment. It's an investment of time and exactly how much time is a personal decision. But zero time will yield zero results.

---

17. http://www.tinyurl.com/34r7ld (http://www.forrester.com/Research/Document/Excerpt/0,7211,45128,00.html)

Don't be pushy

6 Provide really useful information and content

In a seminar I presented to members of the furniture industry at the High Point Market Seminars
in High Point, North Carolina, I shared my thoughts
on what I believe people want from online communities. They want knowledge and new ideas; advice,
acceptance, and approval; information, interaction,
empathy, and support; and purchasing advice and
useful tips for their everyday lives. If you can provide any of that, it would behoove you to give online
communities a try.

We all possess information that is useful to someone else. Determining what that is and delivering it
in bite-size digestible pieces is a great way to gar-
ner and retain interest and engage an online audience. Remember the financial advisors I mentioned
back in Chapter 3? They were sitting on a treasure
trove of useful information that they didn't deem
useful because everyone in their office was well
aware of it and pretty well-versed on the subject
matter. You have to remember that people outside
of your circle and certainly your profession and
hobbies do not have the level of expertise as you,
and could likely learn a great deal if you simply
start sharing what you know.

In an interview with Digital Media strategist Kipp Bodnar,[18] Richard Binhammer, a member of Dell Computer's Communities & Conversations team, detailed some of what Dell is doing in the social media space to reach out to customers, potential customers, and build relationships with bloggers. In the interview he mentioned Dell's TechCenter,[19] a vibrant community of IT Professionals, looking for solutions to their technical issues, news you can use, and tech tip videos for the IT professional and small businessperson, StudioDell,[20] and YouTube. Binhammer characterized the company's social efforts as a whole range of activities that are focused on listening, learning, as well as sharing and connecting with customers. While Dell has an entire team devoted to providing useful information, it is not a requirement for success. The key is to make it interesting as well as useful.

What constitutes useful?

During the week of a winter ice storm, residents of a Seattle neighborhood came together in the comments section of weather-related stories posted on West Seattle Blog,[21] a neighborhood news site created and operated by journalist Tracy Record and her husband Patrick Sand. Record says the comments section of their weather posts turned into incredible neighbor-helping-neighbor discussions with people sharing information on everything from whether the bus was running to where to buy or borrow a snow shovel. Now that's what you call useful information. Not only was it useful but it was timely. The same information would not have been as valuable on a clear, sunny Seattle day. This illustrates the opportunities that exist for anyone who can provide timely, helpful, sought-after information.

During the weeks leading up to April 15th, accountants can hold court in the middle of a busy intersection for an audience of hundreds. Their knowledge is highly sought after during tax time and anyone who can

---

18. http://www.digitalcapitalism.com/
19. http://www.dell.com/techcenter
20. http://www.tinyurl.com/24uyxb (http://www.dell.com/content/topics/topic.aspx/global/shared/corp/media/en/studio_dell?c=us&l=en&s=corp)
21. http://www.westseattleblog.com/blog/

capitalize on that knowledge during a time it is valued most will find success. When a hurricane was expected to bear down on North Carolina, members of my community couldn't get enough information about how to prepare.

Anyone who appeared knowledgeable was in high demand. If you know where all of the deals are during back-to-school season and where school supplies will be marked down significantly, you possess information that parents of school-aged children will find helpful and you should find ways to share it in your community. Strive to repurpose your knowledge into catchy headlines, unique "how-to" and "ten things you didn't know about" articles and blog posts. Find ways to spark interest and fill existing needs. The key is to approach this method with an open mind, identify your areas of expertise or where your knowledge surpasses that of the average person, and distribute that information when it is most valuable. There are varying degrees of usefulness, and the term itself can be somewhat subjective but your content will surely resonate with someone.

One of the most recent displays I've seen of this on GOLO is by the owner of a local company called Nic's Auto Service. What Nic did was set up a group called "Nics Garage" and described it as a forum to "help spread knowledge and tips to the motoring public here in the Triangle." Within that group he displays several image galleries of some of his various car repairs and images of his shop. You can typically find very detailed captions with the pictures with this kind of information:

"This gallery represents work I did to restore the 'tightness' in the front end of this 1998 Durango with 186,000 miles." Even a simple image gallery can be useful. Anyone who happens to peruse this particular image gallery will automatically think of him and the vivid pictures he posted of the process of rebuilding the front end of a Dodge Durango, if they themselves are ever in need of that kind of service. They will also deem him an expert in body work and car repair and come to him for that purpose, or refer others. I have reached out to Nic encouraging him to share even more of his expertise because it is extremely useful and no one else is providing anything like it.

Now, keep in mind that it is very important to identify and encourage users like Nic to continue contributing useful information that benefits the community and to recruit others to do the same. But that does not relinquish you, the person behind the community, from making it a priority of your own. One of the best questions to ask yourself when it's time to create content for the community is: "What can I do that people will find useful, or helpful?" I ask myself that question every day and most days I deliver, with that being the end goal. Sometimes I deliver on a large scale and at other times it's much smaller. I may host a live chat with financial advisors that consists of an hour of free advice one day, and leave a simple comment for a blogger who posted their resume asking advice on how they can make it more effective, the next. You can be useful in many ways. Stay in tune with the community. Look for clues through comments and other posts to find out what it is the community needs, and find ways to provide it.

7 Ask questions

Asking questions is one of the easiest ways to get people talking fast. A good question can go a long way and keep a conversation moving for days. It can be a very effective method for driving engagement when done correctly. But it's the way you pose the questions and the context you build around them that really get the conversations started.

I posted a blog on one of the coldest days of the year, which proved to be the new record low for that particular day, asking: "What was the coldest day of your life?" Within that post I shared my own story with rich details, chronicling a subzero day in Michigan when I was in high school and forgot my earmuffs and gloves and waited for the city bus for more than an hour before finally deciding to walk home.

I recalled the numbness and tingling in my hands and ears and feeling as though both were actually on fire. Not only did people want to share their own accounts after reading mine, but they also expressed empathy for my experience which happened more than 20 years ago. Similar stories poured in, and everyone wanted to talk about the coldest days of their lives, which is exactly what I'd hoped for. It was a simple question, but it had a major impact that day and was the most visited post on the site for two straight days.

I've seen this type of thing snowball on numerous occasions, and while I do it purposely and with a mission in mind, members of the community ask questions for many different reasons. They may be seeking support for a recent action or decision, or they may want to vent about something as simple as a bad customer-service experience and hope to hear from others who have had similar experiences. They may want to lighten the mood with a little comedy and post a fun question like "What would you do if you had a million dollars?" or "What do you hate most about your in-laws?" But most times, the questions are centered on an issue, personal or otherwise, and that's what pulls people in.

A question like "What did you do last night?" with no real context and no interesting story of an amazing, fun-filled night of your own as a basis for the conversation will not have the same effect as any of the three mentioned above. It's just an empty question planted to suck people in. They may fall for it, but it isn't likely that meaningful interaction will follow.

Some of the most thought-provoking questions I answer on the Web are posed by blogger Chris Brogan.[22] He has thousands of readers, followers, and fans because he's a smart proven expert who isn't afraid to share his personal knowledge and encourage others to learn from his success and his mistakes. Brogan's blog is a community in its own right. People come back day after day to read his posts and leave comments which in many cases are more insightful than Brogan's original post, and he'll be the first to say so. He has managed to build a community of smart, savvy professionals who spend inordinate amounts of time commenting on his blog, contributing to its success.

Successful bloggers like Brogan understand that they are managing a community of people who have an interest in their content and expertise and value their opinions and disbursement of information a great deal. The best bloggers acknowledge the members of their communities in several ways. They respond to their comments publicly, send personal e-mails thanking them for their participation, and oftentimes mention them in subsequent posts or even let them guest blog. They read and take interest in user comments and follow up with questions when they want to know more or believe their audience would benefit from more information. One of Brogan's strengths is the questions he poses which are purposely

---

22. http://www.chrisbrogan.com/

posed to make readers think outside the box, question their current beliefs, and push them to want to do better and learn more.

What questions should I ask?

The questions you ask should have everything to do with your mission and the makeup and shared interests of your community. If you're looking for feedback on a new product, ask for it, and be specific about what you want. Don't post: "Do you like the new product" which will draw one word answers, when you can get more direct answers and start a conversation among community members with a question like "What do you love and what do you hate about the new product?" You should always pose questions that will make people want to act immediately. Tie them into current events, or ask questions that will spark a conversation about an issue that everyone typically avoids or finds taboo. As the community manager of a local site where most members are geographically connected, I can focus on almost any local or state issue or hot topic creating a buzz in the community, such as the record cold day mentioned previously, and start an amazing conversation. I can also tap into the shared history of those native to the area and ask questions that spark repeat trips down memory lane.

Everyone loves a trip down memory lane, particularly when they feel times were "better back then." Longtime WRAL-TV news anchor Bill Leslie has built an impressive following on his WRAL.com blog "Carolina Conversations"[23] by asking questions that spark those trips down memory lane. He asks questions about all-time favorite snowstorms, favorite barbeque spots, best cold-weather restaurants and in-state vacations, favorite downtown areas, and even advice on how North Carolina natives pronounce certain words. He asks users to share their nicknames, favorite election of all time, best chili recipes, and memories of loved ones who have passed on.

Questions can kick off the best conversations and those conversations often lead to much more. What it does is illustrate a high regard for input and active participation in the community and that can be very attractive to newcomers, and keep regulars coming back for more. I ask a lot of questions, and so do members of the community I manage. It's part of the culture and helps people connect to other like-minded individuals and gain exposure to new ideas. But don't feel as though you have to create a

---

23. http://www.wral.com/lifestyles/travel/blog/1028421/

treasure trove of trivia. Everyday questions work fine. Here are some of the most popular questions I've posed to the community that have garnered lots of response and very lengthy comments. Some which were written up to 1 year ago are still being answered today.

- Are you afraid of losing your job?
- Are you taking the drought seriously?
- Should we let the financial institutions fail?
- Are law enforcement officers wrongly criticized?
- What's your favorite restaurant?
- How much is too much to pay for a wedding gown?
- Teachers: What are some of your most memorable presents?
- Where were you on 9/11?
- What did you think of Governor Palin's speech?
- What will you buy during the sales tax holiday?
- Are you stressed about the stock market plunge?
- Do you have an HOA (homeowners association) horror story?
- How much money did you spend on gas last week?
- Know of any cool Halloween events? Let's start a list!

Again, there is nothing amazing about these questions. On the surface they are just simple everyday questions that may or may not spark a face-to-face discussion. But on the Web, there is power in questions. Moreover, in an online community environment, where people are there because they enjoy sharing, the right questions can do wonders for engagement and growth.

One-size-fits-all issues don't emerge on a daily basis and even if they do, the interest tends to ebb and flow. But when you strike a chord, you'll know it, and you'll be surprised at the stories that develop and the connections that will happen based on those stories.

Whatever you decide to ask, do it with gusto and be sure to have an idea as to what kind of conversation you're attempting to craft and what kind of response you hope to get when seeking content so that you can strive to get even more the next time and measure your success. And if you don't get the level of response you're looking for, keep asking. If you're persistent with your efforts you'll get what you want. Just don't give up.

8 Use your influence

A former boss and longtime mentor and friend of mine once told me that I didn't understand my own power. I had recently been hired to help change the culture at the newspaper and get print reporters and editors interested in working with our television news partners and producing multimedia content.

In other words, I was charged with leading the troops out of their comfort zone. Many of these troops were senior editors in a corporate environment that lived by the hierarchical chain. So, he wasn't referring to power as it related to my place in the all-important chain, but the power to influence those in it regardless of my position on the chain, or lack thereof. It was influence that he knew I could wield, having hired me for the position. But he also knew the corporate chain could be intimidating to newcomers unaccustomed to such corporate order and rigid processes. He wanted me to know and ultimately understand that my job was to make things happen, but to do it by way of influence.

None of the people on the chain reported to me, so I really had no other choice and I never knew how well it would serve me years later, when I accepted a position with WRAL.com that would require me to build a community and influence people I didn't know, didn't see, and would likely never meet.

You cannot understand the full extent of your role as community manager until you see the way you can influence others and get the results you want just by asking, or merely by making a suggestion. At the same time you have to notice the same ability to influence in others. In any given community, key influencers will always emerge. They are respected by most members and can influence in a way that supports your mission or in a way that does not. Bill Johnston, director of community research at Forum One Networks and editor of the Online Community Report, noted in 2007 that community managers are increasingly expected to know who their lead members are, and what effect their influence has on other community members. It's your job to figure out how to influence these influencers.[24]

Influencers in action

When a member of my community, GOLO, blogged that she and her young son were going to be evicted from their apartment, another member immediately took on the cause and posted a blog telling the community that they needed to come together to raise money to help her. He was able to collect $600 in less than 2 hours all due to his standing in the community. If that isn't influence, I don't know what is! I was pretty amazed at the way it happened because online fund-raising like that can elude even the most Web-savvy nonprofit organizations. But this member had people actually calling him with their credit card numbers, sending funds through PayPal, writing checks, and dropping off cash at his house, that he in turn delivered to the woman in need.

Another member who is a major supporter of the armed forces has encouraged at least 100 other members, myself included, to participate in the Adopt-a-Soldier program. Another member, who has been very verbal about her job loss and subsequent job search created a group called "The Unemployment Line" to offer support for members who have lost their jobs as well. They are supporting one another, offering interview tips and input on resumes. And another member has created what she calls the weekly 'Blog for Hope' where people come together to help others in need and donate items that will help them during these tough economic times. Her efforts resulted in a huge yard sale held on a rainy Saturday morning in

---

24. http://www.tinyurl.com/yssh22 (http://www.onlinecommunityreport.com/archives/250-The-evolving-role-of-the-Community-Manager.html)

the parking lot of a gas station. Members showed up coffee in hands and sold the goods that so many people had donated to the cause. They raised $480 that day, forged new friendships, and were even featured on the 6-o'clock evening news. This group is still going strong. Right now, they are raising money for a woman whose belongings are being held by a moving company because she can't afford to have her belongings placed in storage, and they're already organizing another yard sale. The leader of this group is a serious influencer, with a positive mission and that's to help members of the community.

I often use my influence to encourage members to create new groups or write specific types of blogs. I use it to coax new members into becoming more active in the community and to make connections among members. I recently used my influence to create an unofficial advisory board of the most influential members to serve as my eyes and ears and provide updates on the state of the community and its members as well as to get Wal-Mart to respond to a complaint posted by a member. My influence was alive and well this week when I posted a comment on a blog that was actually a recipe for Cheesy-Potato Soup. I suggested that the member who posted the blog bring a big pot to us at the station. Well, the very next day he did. It was unbelievable and completely unexpected. And that soup turned out to be the best soup I'd ever eaten! I had a coworker take a picture of us and I posted it along with a blog about that soup, and his amazing culinary skills.

In online communities the power of one is alive and well, and it's important to note that the manager, leader, or creator of any given community will have influence just by holding the title. But it's how you use that influence that matters most.

9 Pour on the compliments

Let's face it, flattery will get you everywhere, and a simple acknowledgment can go a long way. If you're a community manager, or working to grow a community, and you have fewer than ten compli- mentary interactions with community members on any given day, you are slacking on the job. On the flip side, if you go well beyond ten interactions, but none are compliments, you're really slacking on the job.

If there's one thing I've learned about managing online communities and building relationships is that it takes much more work than you could ever imagine, and your members are your assets. It's a must that you treat them accordingly and show them that you're glad they've chosen to be there. I will admit that there are similarities between strok- ing a few egos (Chapter 4) and pouring on the com- pliments, but they are not one and the same.

Ego stroking is largely for longtime members or regular contributors. The group that you depend on to keep it all together on days your presence is scarce and those who you know will spend a lot of time in the community, greet new members, and engage wholeheartedly for weeks at a time. Those are the egos you stroke. But you can pay those people compliments as well, and you should definitely look for opportunities to compliment everyone you can. It can be the key to getting new members to stick around and passive members to become active. Here is a random sample of a few compliments I doled out one day in December:

- LOVE your holiday decorations! Beautiful.
- Got the figgy pudding recipe. I placed it on the Holiday page. Thanks for keeping me posted!
- What an excellent idea!
- Our first mascot! How fabulous!
- I am so happy to read this tonight! The goal is to spread some holiday cheer and when I see all of the hard work that members put into their decorations, it makes me want to do better.

Yes, I do this daily, and I mean every word I type. As long as you're sincere your efforts will be greatly appreciated and people will be inclined to stick around and continue to contribute. Here are a few more:

- This is a great conversation piece. Good topic.
- Great post. You might want to add a link to your last blog since it's related.
- Loved the pictures from your garden. When can we expect to see more?
- That recipe looks awesome, was it passed down to you?
- I see you're passionate about the problem of drunk driving, you should consider writing a blog yourself.
- Great idea for a new group. I'm sure you'll get lots of members.

Don't wait for the big things to come up to start complimenting. Look for opportunities everywhere. You can compliment someone for helping another member, taking a survey, participating in a poll, leaving a comment, or starting a forum or group. Compliment photos, writing style, recipes, and interesting thought-provoking comments and blogs. As a community manager, I definitely compliment a lot of people, but I've been on the receiving end of such compliments as well, being a regular reader of numerous blogs and a frequent commenter. When one of my favorite bloggers responds to a comment I left on their post indicating how insightful it was or asks me to share a little more, it gives me great joy and shows me that they value the time I take to read their content. It keeps me coming back, and I employ the same tactics with any community I'm charged with managing. What you'll eventually find is this will become part of the community culture and members of the community will begin complimenting one another. Remember, you set the tone and others will follow. Make it a point to acknowledge good work and loyal visitors. And be genuine. Let people know how much you appreciate them.

10 Know the culture and respect it

When the owner of a company that sells watches uploads 750 pictures, the contents of his entire catalog, and writes a blog that states: "Visit my Web site" which includes no less than ten different links within the blog, that's crossing the line for most online community members who have a vested interest in the site's culture. It is quickly concluded that this person is only there to make a quick sell and if selling is not part of the culture, the consequences will be swift. It isn't uncommon to make a bad entrance into a community, and it typically happens because the new member did not spend any time looking into the community with the goal of learning a little bit about how things work.

Every community has its own distinct culture. It may not be obvious in newer communities but as a community grows, the culture becomes much more apparent, making it much easier to identify and adopt. But even in cases where the culture is quite clear, people will challenge it and join with their own intentions and to meet their own specific needs even if they are not a match for that particular community.

Here are some common pitfalls and early mistakes that can be made by new members, particularly business owners who join online communities without taking time to learn and understand the culture.

- Advertising your company Web site with no additional information about who you are and why you're there. This looks like a drive-by to the active community member and you will be ignored and possibly reported to the site manager.
- Promoting without useful content. If you received a shipment of a new line of sofas that are environmentally friendly, write a blog about the benefits of going green and upgrading your furniture to fit your lifestyle, as opposed to "Hey, come buy these eco-friendly sofas."
- Pushing your business or products every time you post. This will be considered spam, and rightly so. The goal is to contribute something meaningful to the community not to see how much you can "get" without giving anything in return.
- Blatant advertising.

We all have our ideas of what we'd like the culture of our communities to be, but it doesn't always turn out that way. As a community manager you work to create that culture. In an online community culture study with more than seventy-five participants from the software, tech, traditional media, social media, online community, and nonprofit sectors, the Online Community Research Network took a broad look at the factors that influence online community culture, and the steps community managers and strategists take in cultivating, and in some cases influencing, a community's culture. When asked *what steps they'd taken to establish a new community's culture, the following actions were highlighted:*

- Recognizing positive participation
- Soliciting and responding to member feedback
- Communicating with members

I agree that those factors are important for establishing culture and as the leader in the community you hope that the members will follow your lead. But you have to realize that the culture will also evolve on its own, and much of that evolvement will be shaped by the community. So even as you attempt to define the culture, know your own boundaries and respect the culture that the community has created as long as it is not turning into something that is in direct violation of your guidelines.

If you're a twitter user, you know that much of the culture was and continues to be created by people like you, other twitter users. The founders may have created the 140 character limit, which is really the only rule of twitter, but there are many implied rules that are widely followed as well. While no one is forced to adhere to any of them, a large percentage of users do and that's what allows so many a rich experience. When a community follows and enforces its own rules, a community culture is born.

I recently met a realtor at a Tweetup (networking event) who confessed to me that he'd joined the online community I manage and filled his posts with links and blatant advertising, only to be publicly shunned by the masses. He went on to elaborate even further about the backlash he received from the community, which was unforgiving and swift. After we talked a bit more and I explained to him why I thought that happened, he said he wanted to try again and I gave him my blessing. I also gave him some tips that could help him start off on the right track and become a valuable member of the community. All of the tips I provided had everything to do with learning and respecting the culture.

Know the culture and respect it

11 Complain

If you're looking for an easy way to get your community members talking, find something to complain about, and do it with gusto. Whether you complain about a personal issue, a local or regional issue affecting the community, political candidates, gas prices, poor customer service, the morning commute, horrible weather conditions, or the rising cost of ruby red grapefruit juice people will chime in at warp speed to seize the opportunity to commiserate. Remember the old saying, "Misery loves company?" It is alive and well in online community settings. It's human nature to complain, so throw out a topic and let the masses run with it.

Keep in mind, though, that your complaint has to be something that most people can relate to. If you're a billionaire who just lost a million dollars in the stock market, you may not get many people jumping in on that conversation to express their sympathies, unless of course you're participating in a community for billionaires.

I've had great success with complaints and I've noticed that those types of posts from members draw a lot of attention and great conversations. One of my complaints, which turned out to be one of the most popular posts of the summer was about gas prices. With prices well past $4 per gallon and no end in sight, I posted a blog with this headline: "How much do you spend on gas each week?" I shared my own story within the post, complete with a bad attitude toward the rising prices and needless to say everyone was more than happy to express their outrage and the similar drain on their own wallets. It went on for weeks, and people started posting even more blogs on the subject and moving into other topics like fuel-efficient cars, hybrids, and whether or not premium gas is necessary for luxury cars or just another rip-off. So there are added benefits to complaining, and one complaint seems to beget another.

Online communities mirror the real world in the sense that most people simply want to be heard, and this is why members feel comfortable complaining about so many aspects of their lives, from the mundane to the very personal. And given the fact that it all happens while one is protected behind a keyboard and monitor, and, in most cases, under what I call the cloak of anonymity, it makes it so much easier to do. What often happens with conversations based on a complaint is that the tide somehow turns and solutions begin to form. Just as people share personal stories when asked questions, they tend to do the same when a fellow member shares a complaint.

Someone who shares a bad experience at the dentist will receive similar horror stories but will eventually get stories from others about great experiences and maybe even a recommendation to a different office, if it's a local community. Someone complaining about a situation with their mother-in-law might eventually start receiving advice on how to make things better or fix the situation. They may even get questions about the role they play in the bad relationship, possibly causing them to rethink their position. While the authors of many blog posts, comments, and forums centered around complaints are happy to have others commiserate, I believe that in many cases they appreciate the advice that comes along with it and simply belonging to a community where you can vent and be heard, and have your feelings both recognized and validated.

12 Make it personal

From sharing stories about your own life and experiences to communicating via e-mail, the community manager has to be willing to share a portion of themselves with the community, and it can be done in ways that are safe, fun, entertaining, and sure to foster discussion. If you don't show your human side and interact with the people who spend so much of their time in your space, you're not tapping into its potential.

I can go on and on about the importance of being a real person who is not only available to members of the community, but highly approachable and accommodating but I'll show you instead. Here is a list of all of my public exchanges with community members over a two-day period. It does not include the numerous e-mail exchanges over that same period or time spent in the chat area discussing issues and seeking feedback. Some of the comments may be a bit out of context as many times I was responding to a question or request, but the goal is to give you an idea of the level of activity.

- Oh it will be fine! I'll e-mail you when I get in so we can coordinate a good time. Good morning, BTW. I'm making lunches. Always like to check in and see what's happening in the am hours. Later,—Angela…

- Hi. We need to schedule your profile. My calendar is filling up quickly. How's tomorrow?—Angela
- If you're inundated with e-mails—create filters and rules that send them to different folders. That's the only way I stay sane.—Angela
- Tomorrow at 2:30. Send me a number where you can be reached. My e-mail address is: for you to aconnor@cbcnewmedia.com.—Angela
- Hey there GOLO Animal Lovers group! I just posted two new NCSU pet clinical trials in need of dogs with osteoarthritis. Don't know if you know any, but thought I'd pass it along. You can find the two latest here: http://www.wral.com/lifestyles/pets/asset_gallery/2427471/—Angela
- Powerful post today!—Angela
- Welcome to GOLO, Martin. There are a lot of good people here. We care about our members. Please feel free to blog about your feelings or other issues related to what's going on. You'll be surprised by…
- Congrats on that BEAUTIFUL baby! Thanks for sharing on GOLO!
- So you're going to compete with GOLO chat now? Just ask me to open it, mister!
- BTW—thanks to ALL of you who do keep it clean in GOLO chat! I greatly appreciate it. Off to a 10:30 meeting.
- Hey Sandra: What's the latest on your first Blog for hope? Were all the items taken?
- Welcome back Lady!
- Welcome to GOLO! Love that personal statement and that baby is adorable!
- Hi there Arthur! Your blogs will show up on the short list in a few days. You have to be a member for a little longer before your blogs show up on the most recent list on the home page, though they do show up on the longer list you get by selecting "show more." Hope that helps.
- That is a great gallery of Jayda and Creech. I posted it on the WRAL .com pet page.—Angela
- I see you're at the top of the popular list today! Happy New Year.
- Done. My best to your uncle!—Angela
- Hi Gina: Thanks for the compliment! Believe it or not, I do my own hair. I only go to a professional to get it trimmed or cut into a new style.
- NM—Yes. I second that. LOL!

- Want me to call her? J/K! Good luck. You know that GOLO will help you do what's right. I agree with everyone here that you have a good heart.

- Hi there, I want to tell you personally that I removed your blog. Not because of the content for which you are responsible, but for the comments which you are not. You may not see this as fair so that is why I wanted to reach out to you personally. The comments were way out of hand and there is no place for all of that on GOLO.

You may have noticed that my comments were seemingly all over the place with no single theme. That's because each community member is different, has different needs, and a different relationship with me, the community manager. While it isn't always smart to share extremely personal information, such as your address, home phone number, or where your kids go to school, you can still make references to portions of your life away from the community without being a completely open book. I often mention my daughters (though I never use their names), write movie reviews, and share other little nuggets about my own life when it lends itself to the conversation.

When you share, others will share as well. I once posted a picture of myself when the community was about 8 months old and wrote in the headline: "The Real Me." Nearly forty "The Real Me" posts followed, with people sharing pictures of themselves and opening up in a way that I couldn't have imagined. It even spilled on to the next day, and recently made a comeback but this time it was started by a long-time member who clearly wanted some of the newer members to have that same experience. And this was all because I went out on a limb in a community where everyone was pretty much anonymous and put a face with my name and online persona. Share a little bit of yourself with the community. It goes a long way and builds much-needed trust.

Make it personal

13 Seek expert advice and opinions

It's important to recognize the brain power that exists within an online community. In most cases, we have no idea who's on the other side of the monitor or what types of credentials they possess. Unless your community caters to a very small, distinct niche or specialized group it is highly likely that there's a wide range of experience among you, or in other words: a lot of people, who know a lot of stuff.

These experts will emerge in unsuspecting moments. There may be a conversation brewing about a local fire that's been deemed arson and someone chimes in with their opinion disclosing the fact that they were once an arson investigator or that they've been a firefighter for 20 years. On a blog or forum where people are complaining about taxes you may find out that one of your top posters is an accountant with H&R Block or works for the IRS simply by the advice they're offering.

These are bits of information you may not have known previously because there was no reason for the members to mention them before those moments, they simply had not been inclined to do so or perhaps it had not been relevant until then. These are instances that can't be missed and you must make a mental note of the various experts who make up your community to be called on as needed. That kind of intimate knowledge helps build connections among members and will serve you as you strive to bring people together and launch new initiatives.

You can capitalize on this information by asking people to share some of their expertise in a blog or by conducting Q&A sessions with various members, raising their profile and status within the community. These types of displays call attention to the talents and vast amounts of knowledge and experience that exists within the community. If you make it a habit to read through comments and pay close attention to conversations, you will be amazed at whom the real people are behind the online profiles and personas.

But you don't have to sit back and wait for this to happen. You can jump-start the process by initiating discussions of your own, that will make people want to engage and interact. Ask a question directed at experts in a certain field to test the waters. Once you've identified a few experts, reach out to them and see if they'd like to be part of a live chat or panel discussion in their area of expertise. You can also tap into a member's talents by perusing profiles and following comment trails. You'd be surprised at what people divulge about themselves in certain areas, particularly with members they've grown close to and forged real relationships with. Remember, in online communities, conversation is king.

After creating a group in my online community called We are Teachers, I soon learned that there were more than thirty teachers and retired teachers on the site, hence a new pool of experts. Now whenever a new member joins who just so happens to be a teacher, I can send them to that group to meet fellow teachers or ask one of those teachers to befriend the new member and take them under their wing. I can also go to this group for advisement on various topics related to education.

Community manager Krist Klingler[25] takes advantage of the many niche forums she manages for her professional and personal life and is never hesitant to seek input and opinions from known experts who she believes can educate her, and she's always open to their answers. "I'd wanted to

---

25. http://www.kristeclectic.com/

get into video games and didn't know a thing about all the different consoles. So I went to one of my video game communities and asked. I was honest and said I'm an older, first time gamer who feels intimidated by all the options. What should I start with? And they were really awesome about it," says Klingler. "If you've been helpful to them when they've needed it, they'll often be willing to return the favor."

I can certainly attest to that sentiment. When I received a traffic ticket a few short months after moving to the Raleigh area, I took my issue to the community in search of advice. After all, I was in a new state, with new laws and would soon face a very different system without a clue. The community rose to the occasion, informing me of my options and arming me with enough information to make a good decision.

Seek expert advice and opinions

14 Ask for help

As the person responsible for the well-being and growth of the community, it's easy to feel and operate like an island, putting all of the work on your own shoulders. But as the community grows, so does the number of stakeholders. One of the best things you can do for yourself is recognize them as such, build on their sense of ownership, and let them help you build something special.

What I mean by that is look for opportunities to bring the community in on decisions and create partnerships with those who really seem to have a vested interest in its success. Contact your top posters and most involved members and ask them to greet and reach out to new members. Ask them to work on a community-driven FAQ thread. Tell them what kinds of content you'd like to see more of and ask them to help you build it. A simple call for photos or certain types of content often provides the guidance that may be just what some of your less-involved members need to get them more active. Not everyone will jump right in, but you may be pleasantly surprised by the level of response.

It's true that no one should care more about the community than you, the person running it, but as one member shared with me recently, an online community can become a surrogate family for many as real relationships are often forged away from the community. With value like that coming out of communities and transferring to the real world, there is surely a great deal of people who want to see communities thrive and grow.

If you're wondering what kind of help to solicit, take a few days to think about areas where you may be overwhelmed or struggling. Is there something you always say you're going to do but never get to it because of "other things?" Why not take a good look at those "other things" and determine whether or not you can seek help in those areas. Mulling over new features but not quite sure which to implement next? Ask. Put the question out there and let the members help you decide. It would be a shame to put a ton of work into creating some new feature or introducing a new tool only to find out that something else would have served the community much better or filled a need that you may not have been aware of. It would be even more of a shame if you chose that particular feature over the other because you never sought input from the people who would use it.

Receiving such advice can help guide your decisions but seeking it in the first place illustrates a level of inclusiveness that even the most apathetic members can't help but appreciate. I am learning firsthand the power of enlisting a small army of truly vested, caring members as a sounding board and resource. In addition to reaching out to the community as a whole, I've also organized a handpicked group which I use to generate new ideas and offer feedback.

We have a private invitation-only group, with a small number of members who I refer to as the unofficial advisory board. For every idea I bring to the table, they build on and bring ten more. We discuss pros and cons of these ideas, member behavior and trends, and they tell me openly and honestly what works and what doesn't. This small group also enlightens me about how the community is being used by others and what could make it better. They also seem to benefit from a closer connection to me as I often see them defending decisions, taking on troublemakers, and operating like bona fide PR reps for the community. It's pretty amazing to watch. I have learned that people are happy to have a role in the community's growth, so I suggest being just as happy to provide it, because the more evangelists you have the better. They can be your eyes and ears, and capture the pulse of the community in ways that even the most plugged-in community manager cannot. And depending on the tools and features you have at your disposal you can even upgrade the status of your top operatives giving them power to do some light moderating or perform some other function you could use help with.

Lucy McElhinney, community manager of parenting site UKFamily[26] in London asked her members to test a feature on a new sister community and that proved to be a huge success. The result was great member involvement, new registrants to the site, and ace user-testing. When in need of new graphics, Krist Tsirk takes the need to her forum regulars in the form of a design-a-shirt contest, where the winner gets a free shirt, some cash, and a sense of pride knowing other people are enjoying what they made. She believes that any form of patting on the back leaves people feeling good about helping and wanting to do it again. Those are two examples of pinpointing a specific need and reaching out to the community to get that need met.

You may find that it's hard to pinpoint exactly how you need help, but it will become crystal clear in due time. And when it does, ask for help. I'm 99% sure you'll get it, but I'm 100% sure you won't if you never ask.

---

26. http://www.ukfamily.co.uk/index.html

15 Accept and respond to criticism

In a forty-two-page slide posted on her blog and the slide-sharing community, Slideshare, Dawn Foster of Fast Wonder Consulting Company describes in great detail what is expected of a community manager and the many hats we are required to wear.[27] She also shares a list of seven things that she believes make community work. One of them is responding to criticism, and never deleting negative comments. Her advice for dealing with negative comments is to respond constructively. This is not always easy to do, but it is the smart thing to do. The good thing is it will ultimately feel like the right thing to do once you accept the fact that you do not own the community.

The community owns the community and transparency on your part is vital. There are times when I feel like I've spent the better half of my day responding to comments and e-mails and the days when they are largely negative or critical can be especially daunting. But taking the time to respond constructively can yield several positive results:

27. http://www.tinyurl.com/brfukt (http://www.slideshare. net/geekygirldawn/online-community-management-yes-its-really-a-job-presentation)

- You may find that the criticism was legitimate and make a much needed change.

- Your response may address a long or widely held misconception throughout the community that is now cleared up and that can be passed along by the person you were communicating with.

- You will gain the respect of someone who never expected a personal response, and they will spread the word.

- Your reputation will remain intact.

I've resorted to sharing some of the negative and critical e-mails with the community from time to time in a regular blog post called: "What's in my inbox?" and it has become quite popular. The community seems to appreciate getting a glimpse of some of what I deal with and many are often amazed at what people will actually compose and then send via e-mail. One member commented recently that she was going to send me a "virtual aspirin and crying towel" for having to deal with so many levels of criticism, complaints, and sheer ugliness. Another member who operates a home daycare, always comments on those blog posts that my "kids" are worse than hers, and jokes that I should find a way to incorporate naps into their day.

When dealing with these types of comments and feedback I always use discretion and think about how I might feel if my response were made public and perhaps posted on a bulletin board, even though it's a seemingly private e-mail exchange. While it may be pretty obvious to choose your words carefully when responding publicly, it's easy to change your tone and be a bit more relaxed in an e-mail setting but it's even more crucial to choose your words carefully when responding via e-mail or other methods that are perceived as private. It may feel as though you're engaged in a two-way conversation, and you may very well be at the moment, but all it takes is one person to bring many more into the fold. It's best to assume your comments and responses will be forwarded to other members and copied and pasted in many places.

Keep in mind that the community is going to challenge you, as well it should. But you have to be ready for the challenges, respond in a timely fashion, and express gratitude for their concern when warranted. I've often said that community managers or people behind communities must possess an enormous amount of tact, and that now seems like an understatement because it can get pretty ugly at times. But the way you deal with that ugliness will resonate with those who witness it and will serve you well in the long run.

How others respond to negative feedback and criticism

There are several schools of thought when it comes to responding to negative feedback. One approach is to respond to it all. Another is to respond selectively, or on a case-by-case basis. Legal consultant David Marshall says if the criticism is justified, he demonstrates a willingness to change, and if he disagrees, he responds in detail, setting out precisely reasoned explanations and rebuttals. He says he's always friendly even if in what he calls "a distant kind of way." Tomer Lanis of Credit Suisse says an intelligent response to criticism might be able to diffuse it and leverage additional sympathy, integrity, and reliability. He responds to valuable opinions with acknowledgment and gratitude.

Marketing and public relations specialist Angela S. Hwang agrees with Lanis in terms of how a response can certainly diffuse the situation. "You can sway the public's opinion by responding and thoroughly explaining yourself and clearing any misunderstanding so that you won't be viewed negatively." As a PR professional, Hwang says anything that's written about potential clients or people she would represent matters. "It's my job to control the content rather than letting others spread lies." Scott Meis, senior project and social media director at Carolyn Grisko & Associates[28] in Chicago responds to negative feedback and criticism in three ways: openly, honestly, and with transparency. "The conversation about your company, client, product or brand is going to take place no matter what. You're now putting yourself at risk by not engaging in that conversation. You'll be respected for tackling any negative feedback head on and can hopefully listen, learn, adjust and be proactive about making sure the issue doesn't come up again," says Meis who was once assistant public affairs officer for the US Navy.

Patrick O'Keefe, author of the book *Managing Online Forums* says he's "totally happy and ready to address any negative feedback that is at least somewhat respectfully phrased." Brian Whalley, community manager of

---

28. http://www.grisko.com/

OurStage.com[29] gives the same reply for all negative feedback, but depending on the feedback may or may not directly reply to their concern. Negative feedback and criticism is so prevalent on the Web that the US Air Force has coined a special term for how it deals with such feedback: "Counter-blogging."

On David Meerman Scott's blog WeblnkNow,[30] Capt. David Faggard, chief of emerging technology at the Air Force Public Affairs in the Pentagon defines counter-blogging as when airmen "counter the people out there in the blogosphere who have negative opinions about the US government and the air force." They even have an air force blog assessment flow-chart[31] that offers guidelines on how to deal with various types of blogs that fall into four distinct categories and how to assess whether or not a response is in order for each. The categories are troll, rager, misguided, or unhappy customer.

Bryan Person, social media evangelist at Live World agrees with the approach of evaluating whether or not it makes sense to respond to negative comments and posts. Social media marketing consultant Marc Meyer subscribes to that train of thought as well. Meyer's advice is to take the criticism head on, then determine if it's constructive or destructive.

Mortgage banking manager Jerry Gardner agrees with making such assessments. "I think it is important to pick your battles, and therefore respond on the merit of the criticism. Some people criticize simply because they want to see themselves in print, and therefore don't merit a response," says Gardner. Though merit does come into play, Gardner says criticism, in general, should be viewed as a welcome component and provides opportunities to either stand firm on your position by supporting it with further evidence, or correcting a mistake in judgment by way of criticism that proves you wrong. He says in all cases one should "keep to the high ground" when responding and resist the temptation to take the criticism personally.

---

29. http://www.ourstage.com
30. http://www.tinyurl.com/5p6n9r (http://www.webinknow.com/2008/12/the-us-air-force-armed-with-social-media.html)
31. http://www.tinyurl.com/87tud9 (http://freshspot.typepad.com/.a/6a00d83451f23a69e20105365f0d62970b-800wi)

"It's hard for many companies to understand the correlation between listening and response," says John Cass, author of the book *Strategies and Tools for Corporate Blogging*.[32] "They understand customer service but it's entirely different when it happens in the context of the web." Cass believes that part of the reason companies don't engage or otherwise manage such two-way communication is due to a lack of current infrastructure. "Everyone's asking, 'Who is responsible for that?' 'Should we give it to PR or customer service?'" It's certainly understandable not to know where this concept of "engagement" fits within an organization, but it's a mistake to pawn it off in an existing area with no plans to make it a priority. That is not a strategy. It's a recipe for disaster.

---

32. Cass, John. *Strategies and Tools for Corporate Blogging* (Elsevier, Inc. 2007).

Accept and respond to criticism

16 Make small talk

If you have no interest in the nuances of people's everyday lives, you'd better get some, and fast. It's one thing to come up with thought-provoking questions, provide useful content, and encourage community members to interact, but quite another to carve time out of your busy day to initiate small talk with community members. I'm not talking about posting a blog about the weather, current events, or weekend plans, though all three are fine fodder for small talk and widely recognized as conversation starters, but a direct one-to-one conversation between you and individual members of your community.

Greeting new members is the perfect opportunity to initiate a little small talk. Instead of simply welcoming them to the community and leaving it at that, look for clues about them that you can use to create a more personal welcome. Do they list their favorite movies, hobbies, Web sites, books, or other interests? If so, ask them to tell you a little more about any one of them. If the movie *Titanic* is listed as their favorite and you love the movie as well, mention that and ask about their favorite parts of the movie and share yours. Do they love to cook? Inquire about a recipe for a dish you'd like to prepare, or share one of your own.

Make it a habit to respond to comments you come across even if they aren't directed at you specifically. I regularly drop by a member's profile page to let them know I saw their comment on someone else's post and it made me laugh, or made me think, whichever is the case. That always leads them back to me with a response or more details about that specific comment and before you know it, a new connection with a little meaning behind it is made. You make these kinds of connections one by one, but once they're made they stick.

Here's an example: A few weeks ago I stopped in the chat room on the site I manage and a few members were talking about their adult children. I knew from a previous blog that one of the members had an adult son who had recently moved back home and was looking for a job. I was able to join the conversation by asking that member about his son's job search. A conversation ensued and during that back and forth, he mentioned his son's name, among other things, which I hadn't known until then. I made a mental note of the name, so the next time we communicated I could specifically say to him: How's Luke's job search? Did he find anything?

That simple question illustrates two of the 18 rules of community engagement: Make small talk, and make it personal. The more knowledge you have about members, the better your chances of making real connections. Think about how you feel when someone asks about your children and they use their real names. Anyone can ask, "How are the kids?" But someone a bit more connected to you might ask: "What are Johnny and Jane up to these days?" or "How was your children's first day at camp?" See the difference? It's pretty much the same question but inserting the names or mentioning something specific illustrates a higher level of interest and will likely yield a much more in-depth response.

We have to understand the reasons why people join online communities. The reasons may vary from one person to the next, but no one would get involved if they didn't have a strong desire to communicate and connect with others. And much of that communication begins with small talk. I've received e-mails from many members indicating how much they appreciate the friends they've found by being part of the community, and a great number of those online friendships have turned into offline friendships as well.

One woman, who was relatively new to the area, told me during a phone conversation that if not for the community, she wouldn't have had any friends at all. Another told me that she was able to deal with her recent job loss because of the outpouring of support from fellow members. There are times when a simple question like "How's the job search?" or a comment like "Hang in there, something will come along soon" can make all the difference in a person's life.

So whatever you do, don't discount the effect that your small talk can have on others. I am continually surprised at the responses I receive from members after posting a brief comment on their profile page or asking a simple question. I realized the power of these small gestures after noticing a pattern in the responses I'd receive from members after leaving a simple comment or question on their page. That first sentence of many of the responses begins with "Thanks for asking, Angela" or some variation of gratitude for my interest. What that tells me is that people like and appreciate knowing that someone is interested in them on some level, no matter how small.

So going back a bit to those reasons why people join online communities, I strongly believe that two of them are understanding or empathy and support. Is this something you can do on a grand scale, every single day? Probably not, but it is something you can do in some capacity every single day. It's definitely easier to reach a larger percentage of people if your community is small, but don't feel intimidated if you run a very large community. I am not suggesting that you have to engage in small talk with every single member, nor am I saying you have to learn the first names of thousands and the names of their children and grandchildren to be effective. I recognize that there are days when you will certainly be bogged down with administrative tasks and other aspects of your job and cannot spend a great deal of time communicating back and forth with members. Believe me I do. But again, do not discount any method of interaction. Actions will always speak louder than words, and in the online world, our words are our actions.

Make small talk

17 Tune out troublemakers

It would be completely negligent for me to allow you to read this book about engaging online communities without providing some elements of the dark side that are sure to come along with it. In any community, virtual or otherwise, there will always be people with questionable intentions and those who live to wreak havoc. Remember the old saying "Misery loves company?" It's true. And in the online world it has everything to do with what I call the cloak of anonymity. Think about it. Some of the people you deal with daily may be nothing more than a screen name and an e-mail address, and there is no accountability in that. The ease with which people can cause trouble has not been lost on those who enjoy doing it and online communities can be a breeding ground for these types because they thrive on attention.

Unfortunately, communities come with a built-in audience, and that can translate to a great deal of attention. One of the things I realized early on was when it comes to content generated by the community, or people of whom you have no control, the dynamics shift a bit. In most cases you don't know their intentions, and even when you think you do, you could easily be mistaken.

I've seen my share of troublemakers. Some stalk and harass members, while others constantly push the envelope and violate community guidelines to the nth degree. People will conduct themselves in ways that you simply cannot understand, nor should you try. There are people who will provoke, attack, ridicule, incite, and attempt to sabotage the community, causing all kinds of grief along the way. Some will go away if ignored, others will be relentless.

The question then becomes, do you tune them out or take them on? Martin Reed, creator of the online community Female Forum, says the best way of dealing with troublemakers is through communication. That would be taking them on, but in a nonthreatening way. Martin has mentioned to me on numerous occasions that he sees communicating with the disgruntled party as a step in the right direction toward diffusing the situation. Blogger Kristen Tsirk is not averse to communicating with problem members but says that isn't always the answer because you "can't use reason with unreasonable people." Ruby Sinreich, founder and editor of Orange Politics,[33] says she puts people on notice, limits their abilities, and then bans them. "Online communities bring some of the least mature adults out of the woodwork. I often have to spank them on OrangePolitics," she says. Blogger Brian Whaller says troublemakers wielding negativity need to shape up or ship out, "Turn them positive, or show them the door. I don't have the hours in the day to babysit and teach adults how to play nicely."

Mark Oshiro, community manager of BuzzNet,[34] follows a four-step process when dealing with troublemakers: (1) try to reason with them, (2) give them time-outs, (3) remove them if necessary, and (4) ignore them. Oshiro characterized some of his experiences as a community manager in the comments area of one of my blog posts:[35] "I've been the Community Manager for Buzznet for a while now and I can't even begin to list or explain the abuse I've gotten from people. Harassment, hundreds of angry e-mails, death threats, homophobic rants (I'm openly gay on the site), people pretending to be me on the site in order to make people angry at me, attempted hackings of my page ... the list doesn't end there." But

---

33. http://www.orangepolitics.org
34. http://www.buzznet.com/
35. http://www.tinyurl.com/6crhwx (http://blog.angelaconnor.com/2008/09/05/with-community-management-comes-a-new-kind-of-stress/)

Mark doesn't let this type of behavior turn him away. He copes by releasing his rage for 10 seconds "before" responding or reacting online, and by having a sense of humor about the whole thing. Tracy Record of West Seattle Blog[36] doesn't ignore troublemakers. She deletes posts and comments that break the rules but will only ban members as a last resort. It can be tough to distinguish the actions of saboteurs or "trolls" from those who are merely criticizing you or the site and offering feedback that may be a bit negative but valid, and you have to learn the difference. A member of my community once indicated in an e-mail to my boss that my moderating policies had caused her "undue emotional stress." Someone else warned that he would continue to bring a flurry of problems through his posts and purposely disrupt and even attempt to ruin the community if I did not completely remove another member from the community. Another person commented that I must be "sexually repressed" because their blog posts, riddled with sexual innuendo and inappropriate content, had been removed.

If you are going to bring people together online, engage them, and keep them interested, you have to build up a high tolerance for troublemakers and learn how to deal with them. It could be by simply ignoring them, reaching out to them privately via e-mail, suspending their posting privileges for a day or two, or banning them from the community altogether. Only you will know the correct method, and decisions should be made on a case-by-case basis.

Know when to get tough

As community managers, we really do want everyone to peacefully coexist. However, this can't always be the case. In fact, it is rarely the case. As in life, not everyone in a community, real or virtual, will get along. That's just the way it is.

As the leader, charged with growing the community and helping to cultivate relationships, you also have to know how to step in and take action. Sometimes that action means banning members from the community. It's not something you want to do often by any means, but you do need to know when there is simply no other choice.

---

36. http://www.westseattleblog.com/blog/

Here are seven situations that could lend themselves to banning visitors:

1. They continually push the limits and ignore your guidelines or Terms of Service
2. They are being openly defiant as a means of getting attention
3. They are harassing other members on a continual basis with no end in sight
4. They live to post inappropriate links and not much else
5. They are recruiting others to join a destructive cause within your community
6. Everything they post is hostile and an effort to create chaos
7. They are disrespecting or attacking you publicly and making the issue personal

I am not indicating that each of these situations should result in a banning. I've had every single instance occur in my community and I was sometimes able to communicate with the person and reverse the situation, which is ideal for both sides. But I've also been in situations that were utterly hopeless, and banning was the only way.

There is a fine line between building community and destroying it and as long as there are standards to uphold and guidelines to enforce, you will always walk a fine line. You can't feel bad about this, either. It comes with the territory. I recently found myself torn about whether or not to ban a longtime member who had been pushing the envelope and testing the limits for months even though I'd asked him to stop on numerous occasions. Once he began publicly mocking the rules and posting blogs challenging my authority, I had no choice. He later came back using one of several profiles he'd created which were apparently for the sole purpose of creating chaos. You must realize that managing a community comes with tough decisions, and know that the top posters and most visible members of the community aren't always "pillars" within it. You have to do what's best for the majority and the people you want to find the community attractive. You have to learn to be tough because managing online communities can be a real challenge.

18 Showcase and acknowledge good work

If you don't have a way to prominently display and showcase the best content created and submitted by members of your community, you should really consider working it into your overall strategy at some point in the very near future. If you don't think it's an option or if you are not in a position to make changes to the community, then you need to find some way to showcase the best of the best. The fact that you have great content won't mean a thing if only a few people see it, or if it drops out of sight rather quickly.

My community has a home page serving as its front door, with many positions that allow me to give the best content top billing and public acknowledgment while at the same time attract new members who may want the same kind of acknowledgment for their work. But don't be discouraged, or think that you can't provide top billing and public acknowledgment because you don't have a home page or designated space in which to do this. We didn't have this page when the community launched. In fact, we only created it several months after the initial launch, which turned out to be a great idea because it was unexpected, and suddenly, those who had gotten pretty active in the community and had been submitting interesting content found their work being promoted in a new way. There are many creative ways to showcase good content, and you're only limited by your own ideas or lack thereof. Here are a few tactics you can employ to highlight user submissions:

- *Create galleries of similar content:* If you find that several people enjoy posting recipes, create some sort of recipe gallery grouping them all together. This will have several benefits beyond providing prominence for the contributors. It also provides a new resource for the entire community and illustrates an added incentive for participation.

- *Incorporate member submissions into your own content:* When you're searching for an image to use for your own blog or even for one of the aforementioned galleries, why not use an image that belongs to one of your members and provide credit on the post? This gives their image a second life and elevates it a bit in the eyes of other members.

- *Post a weekly or monthly "most popular" list:* Delve into the stats to see which posts got the most attention during the previous week or month and share that information publicly. You can provide a link back to the original post and even mention the author by name. This can be done for every type of content you accept: images, blogs, forums, comments—you name it. You can also use criteria other than stats, and simply create an "editor's picks" or "staff picks" list.

The good thing about the three examples you just read is you don't have to sit back and wait for any of them to happen on their own. With recipe galleries or other similar content you'd like to group together, you can take matters into your own hands and ask people to provide the goods. Sometimes these opportunities will present themselves with little or no work on your part and you can simply cull the content and be on your merry way. But if they don't, it's okay to ask. Remember a few chapters back when we discussed the benefits of asking? You can ask people to submit their favorite recipes, and build that recipe gallery as they start trickling in. Sometimes all it takes is a little incentive, and letting people know that they are contributing to a larger endeavor that will likely receive prominent display is just the incentive needed to jumpstart participation.

19 Don't try to please everyone

I've often referred to managing online communities effectively as working to provide the "ultimate customer service experience," or "customer service on steroids." You should always strive to meet the expectations of your community members by, at the very least, responding to e-mails, answering questions in a timely fashion (even if it's just to say you don't have the answer), explaining guidelines as needed, and making it relatively easy for people to contact you. It's important to understand though that the expectations of your community members can and will be a moving target, and they may not always be realistic or even fair. After all, we are not talking about a homogenous group, nor are we talking about a group with the same wants and needs or even motivations for being part of the community.

Even if your community caters to a very specific niche, motivations for participation will vary widely among members. With that in mind, understand that it will be next to impossible to please everyone. Wait, let me correct that statement. It won't be "next to" impossible, it "will be" impossible to make everyone happy. You can do your best to please the masses and you may very well succeed at that, but that's pretty much the extent of it. And whenever you do please the masses, pat yourself on the back, because it doesn't come easy. There will be very few times when everyone is happy especially when you make changes. I am not indicating that you should rule as a tyrant by ignoring feedback and opinions of members, because bringing stakeholders and vested members into the decision-making process when you can is an important part of growing successful communities.

There will be times when you can put decisions before the community, allowing people to chime in with their thoughts and ideas, and ultimately make a decision that you will honor. That is a major part of illustrating your desire to give the community what it wants. So make it a point to consult with the community on changes and decisions whenever possible, recognizing that sometimes you will have to do what you deem best and learn to deal with those members who may not agree with your decisions.

There will be times when you have to make changes that you know will be wildly unpopular among the masses but you set the tone with the way you communicate those changes, and again, by understanding that no matter what you do, you can never please everyone. You can add a new feature that has been requested for months with the idea that it is something everyone will be happy to have at their fingertips and feel pretty confident about the level of satisfaction it will provide, and still hear from a disgruntled minority who hate the new feature and question you about why you would even consider making such a change. Trust me, this will happen.

With every new feature we've introduced to the community, there has always been a few who complain and I've come to expect it. It comes with the territory. You can never make everyone happy. Take the complaints in stride, respond to everyone with understanding but also make it clear why the change was made and how it contributes to the overall goals of the community. These types of personal interactions often serve a dual purpose. You could possibly win them over with the explanation and deeper understanding of what it is you're trying to accomplish, and you may provide them with a better understanding of the difficulty of your job, working to keep hundreds, thousands, or even tens of thousands interested in your community.

20 | Manage expectations

The way you communicate with members will change as the community grows and matures, and you may have to manage expectations from time to time. The quick e-mail responses I doled out to community members at lightning speed early on set a very high standard with those who were at the receiving end of such fast replies. It was both good and bad. On one hand, it showed that I was responsive and available and cared about their needs. Some never needed me in that way again and left with a great experience. Others began to treat every need as urgent and subsequently e-mailed me directly several times each day expecting the same speedy response regardless of the importance of the issue or concern. After getting such top-notch service, some became very critical and impatient when I could no longer provide it, at least not at the lightning speed of the early days.

Managing more than 11,000 members is much different than managing a few hundred and I am realizing that more and more each day as new members come on board. What you can learn from my experience is that it's critical to manage expectations. This is something I did not establish at the beginning, and since I did create such high expectations, I often feel stressed to maintain that level, which I am working on changing. I've even had to stop myself from responding to e-mails on the weekends because I always have my BlackBerry with me making it fairly easy to respond to a small question or concern if I'm sitting in the lobby of my daughter's dance studio or am otherwise available. I don't typically have the time to do that regularly on weekends but it feels like a disservice not to respond when it won't take much effort on my part, even though I'm not officially "on the clock." Part of me recognizes that as a strong commitment to the community, which I definitely have. But another part of me recognizes it as me being a bit of a workaholic, and allowing expectations to skyrocket. Again, I am working on that and looking for a good balance.

One way to manage expectations would be by prominently posting clear guidelines as to what people can expect. For instance: If they send an e-mail, how soon can they expect a response? 24 hours? 2–3 days? Whatever the wait period may be, it's good to establish it on your own. Doing so will set the expectations but also open up opportunities for you to exceed them, and go the extra mile whenever possible. When you do go the extra mile, it does not go unnoticed. You will find that what seems like a small gesture to you can be a pretty big deal to someone else. Every time you receive an e-mail or request from a member, you are faced with an opportunity to exceed their expectations, and it's good to do it when you can, but having those clear guidelines takes away the added stress of giving everyone the VIP treatment, and doing it NOW.

The best thing you can do is aim high, understanding that you will sometimes fall short for various reasons. It may not happen much at all if you have a large staff, but if you are a solo act, maintaining Superman or Superwoman status will be hard to do. Look for opportunities to take something small a step further to illustrate your commitment to the community and its members. I recently supported a yard sale being hosted by some members of my community by donating two huge bags of toddler clothing. It was small to me, but that donation was part of an initiative that raised money for needy families and it showed my support for what was an immense undertaking.

So look for opportunities to show that you are more than just the person focused on attracting more members and eyeballs to your site. Get involved when you can, but don't make apologies when you can't unless it's something you promised to do and can't deliver for whatever reason. You will not be able to help everyone, or personally support every initiative, and it gets tougher as your community grows. But no matter how big it gets, you should always think of it as small so that you won't lose the drive to keep pushing.

21 Realize your work is never done

Starting and keeping conversations going in an online environment is not a part-time job, unless of course you want a part-time community. If you're fine with occasional interaction, sporadic content creation, and a community that lies dormant most days, then I suppose you may not have to put in an enormous amount of effort. Maintaining a popular destination with a flow of fresh content requires an amazing commitment. I would love to tell you that you will get out of a community exactly what you put into it, but that isn't true. You get less, and I can't imagine that changing. That is why your work is never done. If occasional interaction is what you're seeking from a community, the level of your involvement cannot mirror that.

The thing about occasional interaction is it will ultimately dwindle to sporadic interaction and before you know it, the conversations will cease altogether. The content will dry up, submissions will wane, and those who remain will soon pack it up and head out for good. It takes time to gain momentum with any community and a heavy investment is required to get it. The level of the investment may decrease as the community matures, but not by too much, and only after you have a well-established community with loyal visitors who deem themselves stakeholders in its success.

The worst thing you can do even after you have a high number of stake-holders and loyal visitors is assume they will be there tomorrow.

Complacency is a detriment to any community manager. I can recall a 2-week period where I was extremely busy and didn't spend much time at all greeting new members. When I went back to their profiles I was pleased to see that they had all received warm greetings from other, longtime members. I'll admit that I breathed a sigh of relief and felt good about the level of ownership those greetings illustrated on behalf of the regulars and how my work in the community helped build that ownership. But it also made me realize that had they not done that, there would have been a substantial number of new members floating in the wind, if they had indeed never visited the community much before joining, were unaware of how things worked, or weren't exactly sure how to get involved.

It is not realistic for me to greet every single member on the very day they join (outside of the automated e-mail welcome they receive after signing up) but I can certainly make it a goal to communicate with a very large percentage of new members within their first week or two, even if only to say "hello, and welcome." That is what I've learned about my particular community. But I also recognize that what works for one person managing one type of community may not work for another. What's important is that you have clear goals and a clear idea of the kinds of experiences you'd like to provide. You also have to be painfully aware of the fact that there are millions of other places your members can choose to spend their time. You have to continue to win people over and illustrate time and time again why yours is the place to be.

It is my sincere hope that you will take several of the 18 rules of community engagement presented in this book and apply them to the way you man-age your own communities, or those you will manage in the future. My goal here was not to provide a one-size-fits-all solution, because there isn't one, and it's important to know that going in. My goal was simply to give you ideas, encourage you to take this work very seriously, and help you understand that it is not a science, but an art. What I hope this book has done for you is provide a group of tools that you can put into your toolbox and access as needed. So take what works for you, but hold on to the rest in the event that it will one day serve you well.

About the Author

Angela Connor is a multimedia journalist and community manager with a passion for online communities and social media. She is the Managing Editor of User-Generated Content at WRAL.com where she launched and currently manages the top-rated news organization's first online community, GOLO.com which has grown to more than 11,000 members in 18 months.

Angela's news management experience spans broadcast, print and online news at TV stations and newspapers in Cleveland, Tampa, West Palm Beach and Fort Lauderdale. She writes the highly-read blog, Online Community Strategist and is often requested to share her social media insight and discuss the benefits of online communities at conferences and organizational events.

Angela develops policies and best-practices for handling user-generated content and driving user engagement, and serves on the Digital Media Committee for the Society of Professional Journalists. She lives in Holly Springs, NC with her husband and two young daughters.

Recommended Happy About® Books

Purchase these books at Happy About
http://www.happyabout.info
or at other online and physical bookstores.

JASON ALBA
FOREWORD BY BOB BURG

*I'm on LinkedIn
Now What (second edition)???*

This book explains the different
benefits of using LinkedIn
and recommends best
practices so that you can
get the most out of it.

Paperback: $19.95
eBook (pdf): $14.95

HAPPY ABOUT®
AN EXTRA HOUR
EVERY DAY

NICOLAS SOERGEL

*Happy About an Extra Hour
Every Day*

This book provides 300 practical
time management tips and
the tips that are easy
to implement are marked
as "quick wins".

Paperback: $19.95
eBook (pdf): $14.95

42 Rules™ of Social Media for Business

This book teaches readers why social media is important to their business and how they can maximize their social media effectiveness.

Paperback: $19.95
ebook (pdf): $14.95

DAVID COLEMAN

42 Rules™ for Successful Collaboration

This book helps the readers to walk away with a much better idea on how to be successful in their interactions with others via the computer.

Paperback: $19.95
ebook (pdf): $14.95